WILD WILD WORLD

GRASSHOPPERS

by Liza Jacobs

BLACKBIRCH®
PRESS

THOMSON

GALE

San Diego • Detroit • New York • San Francisco • Cleveland • New Haven, Conn. • Waterville, Maine • London • Munich

THOMSON

✦ ™

GALE

For more information, contact
The Gale Group, Inc.
27500 Drake Rd.
Farmington Hills, MI 48331-3535
Or you can visit our Internet site at http://www.gale.com

Photographs © 2001 by Lee Wen-Kuei

Cover Photograph © Corel Corporation

© 2001 by Chin-Chin Publications Ltd.

No. 274-1, Sec.1 Ho-Ping E. Rd., Taipei, Taiwan, R.O.C.
Tel: 886-2-2363-3486 Fax: 886-2-2363-6081

LIBRARY OF CONGRESS CATALOGING-IN-PUBLICATION DATA

Jacobs, Liza.
 Grasshoppers / by Liza Jacobs.
   v. cm. — (Wild wild world)
Includes bibliographical references.
Contents: About grasshoppers — Grasshopper food — Chirping and "singing" — Staying out of danger.
 ISBN 1-4103-0051-X (hardback : alk. paper)
 1. Grasshoppers—Juvenile literature. [1. Grasshoppers.] I. Title. II. Series.

 QL508.A2J23 2003
 595.7'26—dc21   2003001469

Printed in Taiwan
10 9 8 7 6 5 4 3 2 1

Table of Contents

About Grasshoppers

There are nearly 10,000 different kinds of grasshoppers. Short-horned grasshoppers have short antennae, while long-horned have long antennae.

All grasshoppers have strong jaws and sensors near their mouths called palpi.

Grasshoppers are insects. Like other insects, grasshoppers have compound eyes.Compound eyes are made up of hundreds of tiny eyes and give a grasshopper excellent eyesight.

Like all insects, grasshoppers have six jointed legs and a body made up of three parts: the head, the thorax (midsection), and abdomen (rear section). Most grasshoppers have two pairs of wings.

Grasshoppers hear with a membrane called a tympanum. A short-horned grasshopper's tympanums are on either side of its abdomen. Like crickets, a long-horned grasshopper's tympanums are on its front legs. Females lay eggs through the ovipositor found at the end of their abdomen.

Compound Eye

Palpi

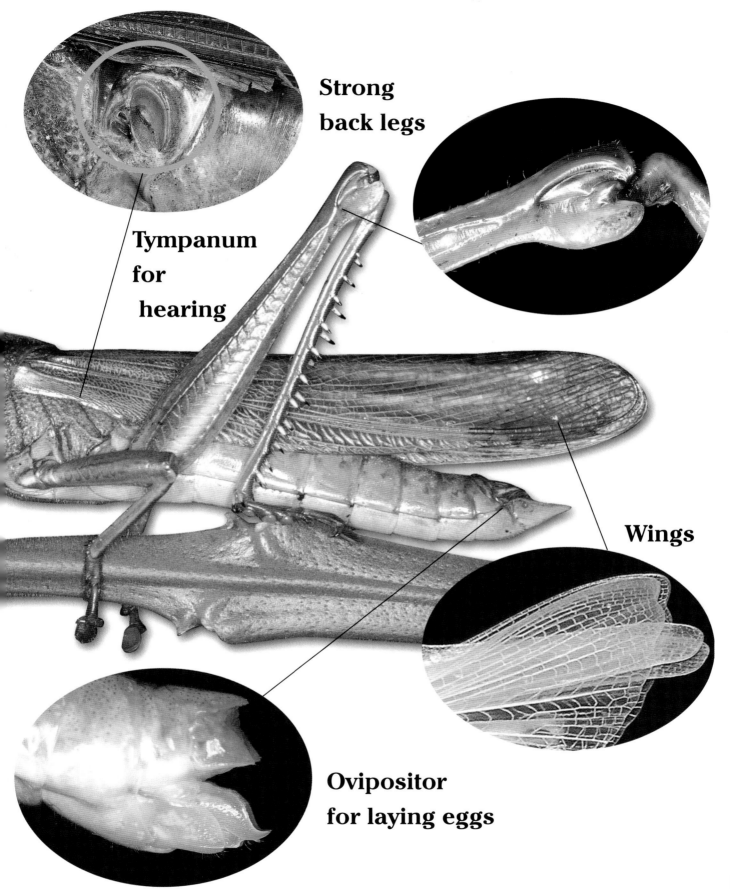

Strong
back legs

Tympanum
for
hearing

Wings

Ovipositor
for laying eggs

Grasshopper Food

A grasshopper uses all its legs for walking. Its long back legs are made for jumping, but the two shorter pairs in the front are used for holding on to food.

Grasshoppers mainly eat leaves, flowers, berries, and other plant foods. Some also eat insects. A grasshopper uses its powerful jaws and sharp teeth to crunch up its food. Its palpi help a grasshopper taste food before eating it.

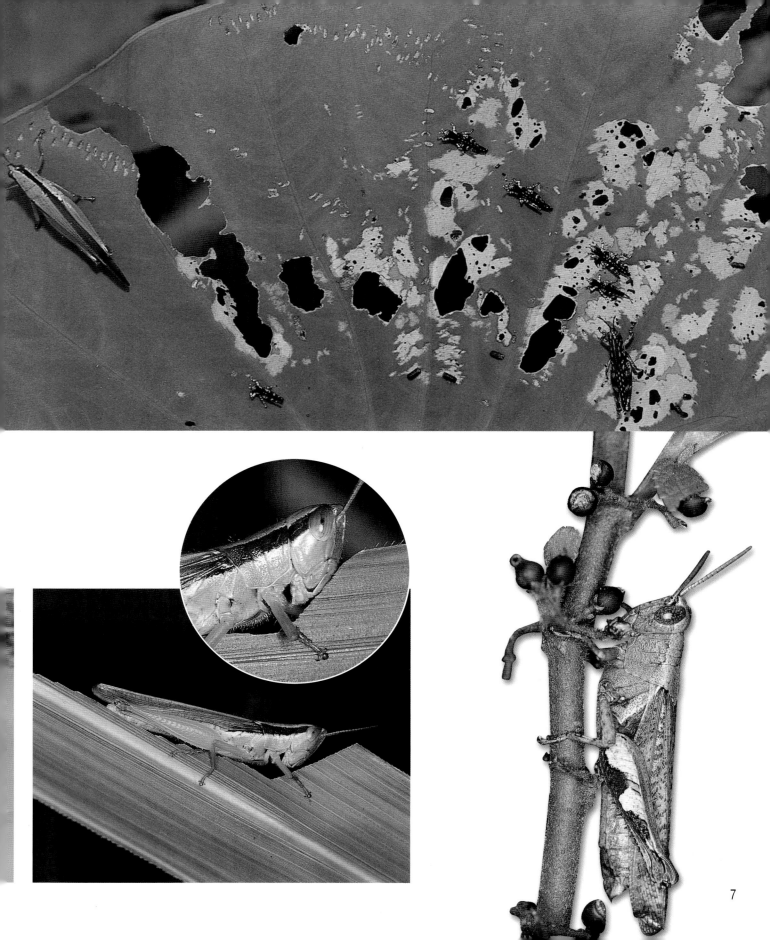

Blending Into the Background

Grasshoppers rest on bark, branches, leaves, and other plant material. They often bask in the sun for several hours at a time. Many kinds of grasshoppers are hard to see while resting. Their shape or color blends into the background. This helps them stay out of danger.

Chirping and "Singing"

Like crickets, grasshoppers make a chirping sound. Some female grasshoppers can make faint sounds, but mainly only males chirp. They chirp to send a warning to another male or to attract a female.

A short-horned grasshopper, such as the one shown here, has hard, ridge-like veins on its wings. It also has small, sharp points that poke out of its hind legs. To make the chirping sound, a short-horned grasshopper rubs its back leg against its wing. Long-horned grasshoppers "sing" by rubbing the underside of one wing against the upper side of the other wing.

Finding a Mate

To find a mate, male grasshoppers chirp. They may also flash the patterns or colors of their wings to attract attention.

The males are often smaller than the females. To mate, the male climbs onto the back of the female. They may stay this way for several hours.

Laying Eggs

A female grasshopper may lay just a few—or more than 100—eggs at one time. After mating, she finds a good place to lay her eggs. She bends and pushes her abdomen down into the soil, where she deposits her eggs. Then the grasshopper coats the eggs with a sticky substance. The cluster of eggs hardens into a mass, which helps protect them from weather and animals. Some kinds of grasshoppers also lay eggs on plants.

Eggs and Nymphs

Grasshoppers go through three stages of life as they become adults. The first stage is the egg. Most grasshoppers stay in this stage through the winter. Then a baby, or nymph, hatches from the egg and digs its way to the surface. Nymphs
are pale, but their color darkens over time. They look a lot like adult grasshoppers, but they do not have fully formed wings.

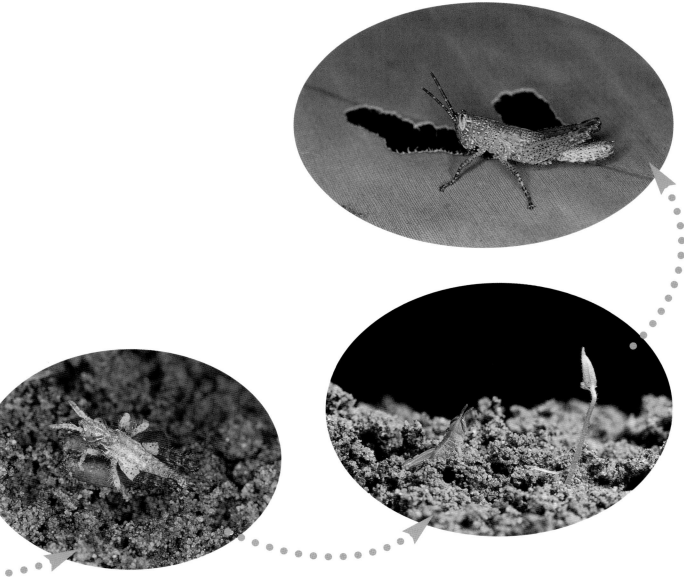

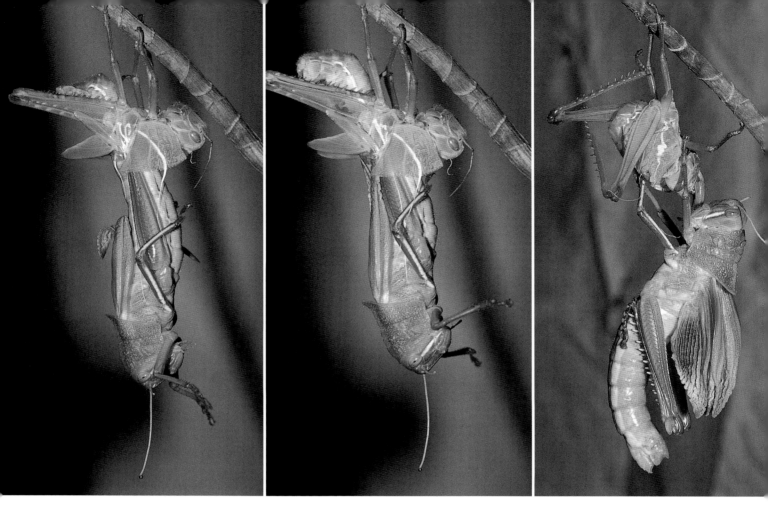

Molting

Insects have a hard covering over their bodies called an exoskeleton. The more nymphs eat, the more they grow. Like other insects, a grasshopper's exoskeleton does not get bigger as the insect grows. In order to grow to its adult size, a grasshopper has to shed its covering 5 or 6 times. This is called molting.

The old skin splits open and the grasshopper wriggles out of its shell with a new, larger covering! Its new exoskeleton is soft at first, but it quickly hardens. After its last molt, a grasshopper has reached the third stage. It is an adult.

Staying Out of Danger

Many kinds of animals eat grasshoppers, including birds, lizards, spiders, beetles, praying mantises, and other insects. A grasshopper has two main ways to stay out of danger. It can stay perfectly still and rely on its brown or green color to blend into the background. It can also use its powerful back legs to jump more than 2 feet to safety!

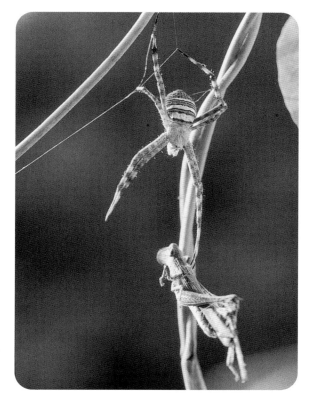

Many Kinds of Grasshoppers

Grasshoppers are closely related to crickets and katydids. There are thousands of kinds—or species—of grasshoppers on the planet. They are found with many different markings and colorings. They come in many shapes and sizes. And they can be found almost anywhere in the world, except in cold polar areas.

For More Information

Allen, Judy. *Are You a Grasshopper?* New York: Larousse Kingfisher Chambers, 2002.

Hartley, Karen. *Grasshopper (Bug Books)*. Crystal Lake, IL: Heineman Library, 1999.

Pascoe, Elaine. *Crickets and Grasshoppers (Nature Close-Up)*. San Diego, CA: Blackbirch Press, 1998.

Glossary

exoskeleton the hard covering on the outside of an insect's body

molt to shed the outer skin or covering

nymph the second stage in a grasshopper's life

ovipositor the reproductive organ through which a grasshopper lays her eggs

palpi small feelers near a grasshopper's mouth

tympanum ear membrane of a grasshopper